What's It Like in Space?

What's It Like in Space?

STORIES FROM ASTRONAUTS WHO'VE BEEN THERE

Ariel Waldman

Illustrations by Brian Standeford

CHRONICLE BOOKS

SAN FRANCISCO

Library of Congress Cataloging-in-Publication
Data

Waldman, Ariel, author.
  What's it like in space? : stories from
astronauts who've been there / Ariel Waldman ;
illustrations by Brian Standeford.
     pages cm
  ISBN 978-1-4521-4476-4
1. Life support systems (Space environment)—
Anecdotes. 2. Space flight—Physiological
effect—Anecdotes. 3. Astronautics—Anecdotes.
4. Astronauts—Anecdotes. I. Standeford, Brian,
illustrator. II. Title. III. Title: What is it like in space.

  TL1500.W35 2016
  629.45'00922—dc23

                    2015018968

Manufactured in China

FSC
www.fsc.org

MIX
Paper from
responsible sources
FSC™ C008047

Illustrations by **Brian Standeford**
Design and additional illustration by **Neil Egan**

10 9 8 7 6 5 4 3 2 1

Chronicle Books LLC
680 Second Street
San Francisco, CA 94107
www.chroniclebooks.com

Chronicle books and gifts are available at
special quantity discounts to corporations,
professional associations, literacy programs, and
other organizations. For details and discount
information, please contact our premiums
department at corporatesales@chroniclebooks.
com or at 1-800-759-0190.

Introduction

"If somebody'd said before the flight, 'Are you going to get carried away looking at the Earth from the Moon?' I would have [said], 'No, no way.' But yet when I first looked back at the Earth, standing on the Moon, I cried."—Alan Shepard

What's it like in space? It's something many of us have wondered about, and something, of course, that astronauts are asked all the time. Much like space exploration itself, the question is boundless and hopeful. Asking "what's it like in space?" represents our collective aspiration to dare how far humans can go and what we can achieve when we get there.

One of the first people to attempt to scientifically answer the question was the astronomer Johannes Kepler. In the early seventeenth century, *three centuries* before a human landed on the Moon, Kepler wrote *Somnium*, a work of science-based fiction about humans in space. *Somnium* detailed how humans could conduct science in space and how the Earth could be observed from the viewpoint of the Moon. The first Moon landing was still

361 years into the future at the time of Kepler's writing, but he was one of the first to declare it scientifically possible and to imagine what it would be like to travel there.

Despite hundreds of years of scientific speculation, no one really knew what space would be like. Leading up to humans taking to spaceflight in the early 1960s, doctors were unsure about everything—from whether people would physically be able to eat in space to whether their eyes would float freely around in their heads. During the Cold War, astronauts and cosmonauts alike were subjected to intense and thorough tests to ensure that they would be able to handle every scenario imaginable. It was only by pushing the boundary of how far humans could go that we could truly answer the question of what it's like in space.

In my work, I make space exploration more accessible. Working on this book has been a fascinating opportunity to learn more about human spaceflight—an aspect of space exploration that has only been experienced by a select few over the fifty-plus years since its inception. But it was mostly just a lot of fun to talk to astronauts who generously shared their time and enthusiasm. The stories range from the comical to the peculiar to the awe-inspiring, but in researching or interviewing each astronaut they all shared a similar undercurrent: a quiet determination to tackle the unknown.

Despite several decades of human spaceflight, any answers to "what's it like in space?" are provisional and will continue to change as the exploration widens, both in scope and accessibility. I enjoy reading through these stories and meditating on what someone four hundred years in the future might write about what it's like in space. I can only hope that space exploration will slowly, but surely, become more accessible over the coming decades and centuries. Maybe someday this book will be as quaint as books describing what it's like to fly in an airplane.

NO ONE CAN HEAR YOU BURP

In space, no one can hear you burp. Seriously. The lack of gravity means you can't keep your food down in your stomach if you try to expel gas out through your mouth. Thus most attempts to burp would actually result in vomiting. Needless to say, this is why there is no soda in space.

MYSTERIOUS HEADACHES

Some of the first astronauts to use the space shuttle reported experiencing mysterious headaches while in space. Lots of money and time were spent researching why. Intracranial pressure? Insufficient oxygen? No. Coffee has to be freeze-dried before being transported to space. This process reduces coffee's caffeine levels so significantly that the astronauts' headaches were actually a symptom of caffeine withdrawal.

MOON FACE

"If you ever go to space, make sure you go for longer than four days," advises four-time space shuttle astronaut Jim Newman. Within the first few hours of being in space, you get what astronauts call "Moon Face." Due to the lack of gravity, your body can't keep the flow of blood as well-distributed below your head. For the first few days in space, your face becomes bloated until your body figures out how to properly distribute blood in microgravity. Generally after the fourth day your face returns to normal and you can more comfortably enjoy your space travel.

LEAKY SUITS

The early male astronauts often had leaky spacesuits. They would frequently complain about their urine leaking into other areas of the suit. For a while, no one could figure out what was wrong with the spacesuits. NASA eventually realized the leaking was due to the oversized condom catheters the astronauts were using. Turns out that when the astronauts were asked by doctors what size they needed, they would often ask for "large."

URINE ICICLE

The space shuttle used a venting system to expel astronauts' liquid waste away from it and into space. In 1984, this system broke down. A huge icicle of urine formed in space, attached to the base of the shuttle. Fearing that the icicle could do damage to the spacecraft, the astronauts had to use a robotic arm to snap it off.

BACKWARDS DREAMS

Astronauts sometimes experience "backwards dreams" while sleeping in space. "Day 52 in space. Had my first backwards dream last night—got back to Earth and gravity wasn't normal," tweeted Reid Wiseman during his stay on the International Space Station. After returning to Earth, Wiseman reported that he still had dreams in which Earth's gravity was abnormal, but that they faded after a week of being back on the ground.

FART PROPULSION

Because it's difficult to burp in space, you fart more. Astronauts have admitted to attempting to use their farts as a type of personal propulsion for getting around the space shuttle and International Space Station. Alas, though perhaps to the relief of fellow astronauts, it turns out that farts don't propel a human body in space.

BARF BAG BOUNCE

Vomiting in space is like a slap in the face. Literally. Without gravity, your vomit will bounce off of the sides of a barf bag and into your face. Astronauts recommend planning ahead by taking a towel with you to clean up. So *The Hitchhiker's Guide to the Galaxy* was right—a towel is indeed about the most massively useful thing an interstellar traveler can have.

THIS IS THE MOON

Astronauts sometimes have to remind themselves of the magnitude of what they are doing. Alan Bean, an Apollo 12 astronaut who walked on the Moon, recounted, "I would look down and say, 'This is the Moon, this is the Moon,' and I would look up and say, 'That's the Earth, that's the Earth,' in my head. So, it was science fiction to us even as we were doing it."

FALLING ASLEEP

Sleeping in space can be difficult. With no bed to lie in due to the lack of our familiar gravity, astronauts have to adapt to sleeping in mid-air by relaxing their muscles enough to drift off. This can be tricky in a floating environment—many space newbies attest to being jolted awake by the feeling of falling, giving new meaning to the term *falling asleep*. One Russian cosmonaut became such a pro at sleeping in space that he was often seen outside of his sleeping cabin, drifting by in a deep sleep, his body occasionally bouncing off the walls.

STUCK MID-AIR

The interior of the International Space Station is covered in handrails, which astronauts use to help themselves move around. To test whether it was possible to move yourself without pushing off of a wall, two astronauts who had flown up on the Space Shuttle *Endeavour* carefully positioned one of their crewmates, astronaut Nancy Currie, in a space where she couldn't reach any walls. Currie found that no matter how vigorously she moved or how hard she flapped her arms, she was stuck in mid-air, unable to reach any of the walls.

WHICH WAY IS UP?

In space, you need to decide which way is up. Unlike the spaceships of science fiction, the space stations of reality use every available wall for storage: no walkways, no ceilings, no empty corridors. Orbiting astronauts find that they have to consciously choose which direction they want to treat as "up" when working on various tasks such as fixing broken items, conducting experiments, and even meeting with fellow astronauts.

STELLAR CITIES

At night, it can be difficult to distinguish Earth from the stars. As astronaut Sally Ride observed:

Part of every orbit takes us to the dark side of the planet. In space, night is very, very black—but that doesn't mean there's nothing to look at. The lights of cities sparkle; on nights when there was no moon, it was difficult for me to tell the Earth from the sky—the twinkling lights could be stars or they could be small cities.

UPSIDE DOWN

Astronauts have been known to perform their own personal space experiments, testing the limits of their bodies in unanchored environments. From peeing "upside down" to blindfolding themselves to see if they get disoriented (spoiler: they do), astronauts are often inspired to explore space with a childlike wonder.

THE SMELL OF SPACE

No one can agree what space smells like. While astronauts can't smell space directly through spacesuits, they have tried to explain the smell that lingers in the airlock after conducting a spacewalk. Answers have ranged from "wet clothes after rolling around in snow," a "burnt almond cookie," "sweet-smelling welding fumes," "ozone," "burnt gunpowder," "fried steak," and a "mild version of the smell of an overheating car engine." The source of all this smell debate lies within the dying stars that created our solar system and produced "polycyclic aromatic hydrocarbons" that cling to astronauts' spacesuits during spacewalks.

OH YEAH, GRAVITY

A common experience among astronauts after returning to Earth from space is dropping various household items, expecting that they would float in place. Some astronauts have tried throwing objects to people, only to watch them fall short before realizing the gravity on Earth requires more effort. Cups, toothpaste tubes, pizza boxes, and pens have all been casualties of gravity forgetfulness.

SPACE ADAPTATION SYNDROME

The majority of astronauts throw up in space. This is attributed to a condition known as space adaptation syndrome, where the disorientating nature of microgravity causes headaches, nausea, and vomiting for the first few days. The worst case of space adaptation syndrome was experienced by U.S. Senator Jake Garn, who became a space shuttle astronaut in 1985. An informal "vomit scale" was invented based on Garn's experience, as astronaut trainer Robert E. Stevenson explained:

Jake Garn, he has made a mark in the Astronaut Corps because he represents the maximum level of space sickness that anyone can ever attain, and so the mark of being totally sick and totally incompetent is one Garn. Most guys will get maybe to a tenth [of a] Garn, if that high.

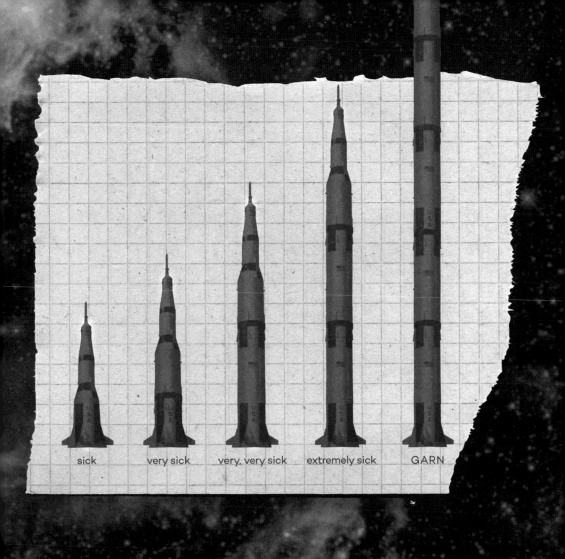

sick   very sick   very, very sick   extremely sick   GARN

SNEEZING IN SPACE

Part of astronaut training includes being taught how to sneeze. On a spacewalk, astronauts are unable to clear away mundane annoyances such as sweat, snot, or tears from their faces, all of which could temporarily blind them. Aside from trying to keep cool and not cry in space, astronauts become experts in aiming their sneezes downward to avoid making a mess on the inside of their helmets.

LIGHTNING

Earth treats astronauts to an epic display of natural fireworks: lightning storms at night. As astronaut Sally Ride recounted:

Of all the sights from orbit, the most spectacular may be the magnificent displays of lightning that ignite the clouds at night. On Earth, we see lightning from below the clouds; in orbit, we see it from above. Bolts of lightning are diffused by the clouds into bursting balls of light. Sometimes, when a storm extends hundreds of miles, it looks like a transcontinental brigade is tossing fireworks from cloud to cloud.

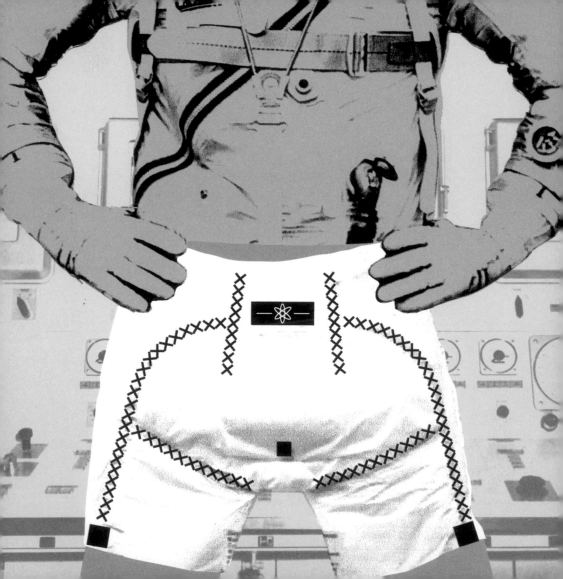

COLLECTION CONTRAPTIONS

Before diapers became standard astronaut attire for launch, landing, and spacewalks, many other options were considered and used. Mercury astronaut Gus Grissom wore a double layer of rubber pants for his suborbital spaceflight in 1961 in order to contain all his urine in a protected reservoir. Other early spaceflights made use of modified condoms attached to urination collection bags. Decades later, in the 1980s, diapers were introduced to accommodate female astronauts. Male astronauts eventually also adopted the use of diapers after realizing that it was a more comfortable way to "go" than the condom-based urine collection devices.

SPACE SILENCE

To the human ear, space is silent, but an astronaut's experience of space is anything but. Between the never-ending drone of space station ventilation systems to the buzz of life support systems inside a spacesuit, space travel isn't the most tranquil aural experience. But while some might consider this to be annoying, astronaut Scott Parazynski viewed it as a type of security blanket:

When you're in your spacesuit, actually, there's a very comforting hum. You're always hearing the purr of your fan, which is your lifeline, it's what circulates your oxygen through your suit. You're hearing the crackle of the radio. It's just kind of a reassurance that you're still connected. When your spacesuit goes completely silent, that's a really, really bad day. So we don't like it to be completely silent.

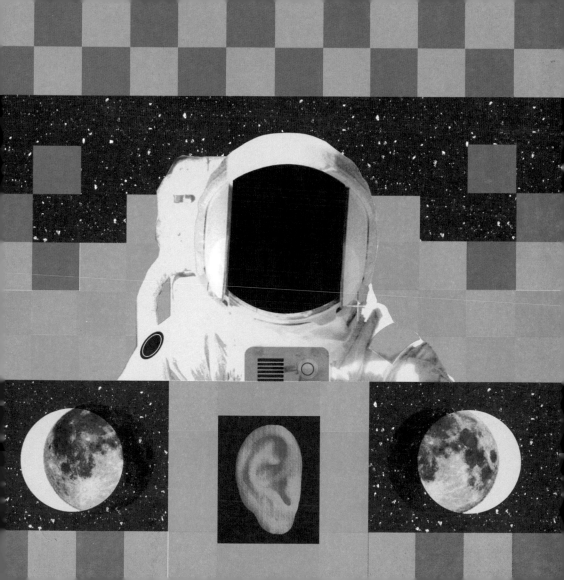

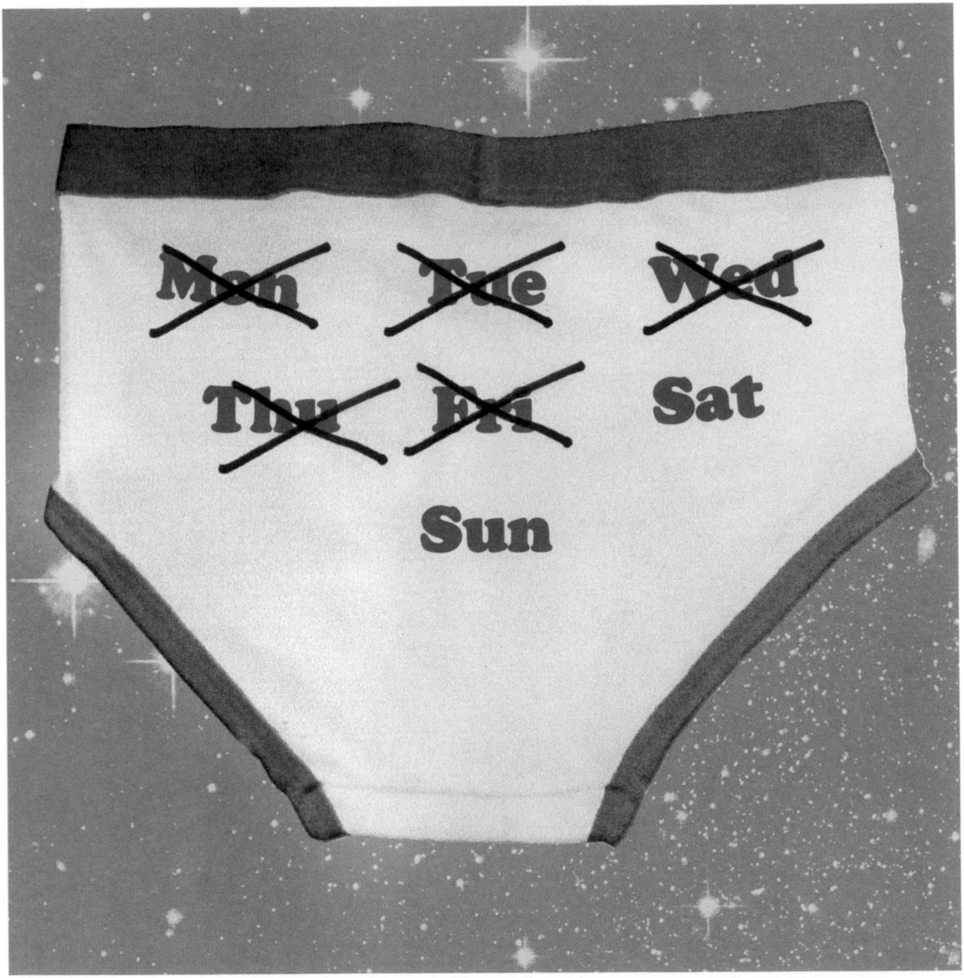
Mon
Tue
Wed
Thu
Fri
Sat
Sun

LAUNDRY IN SPACE

There is no laundry in space. Thankfully, because clothes actually float away from your body in microgravity, they don't absorb all the "funk" they would on Earth. As a result, astronauts have admitted to sometimes wearing the same underwear for several days straight. It doesn't hurt that many people also report losing most of their sense of smell while in space.

STUMBLING HOME

Traveling back to Earth can be harder on the human body than traveling to space. After being in space for a significant amount of time, astronauts' bones and muscles atrophy from the lack of gravity. While this is a serious problem that astronauts work hard to combat, it can result in some funny situations at home. Canadian astronaut Chris Hadfield remarked:

I wouldn't have been able to pass a sobriety test for a week after I returned, and it was four months before I could run properly. In that first week, you're lumbering around like a guy in a Godzilla costume.

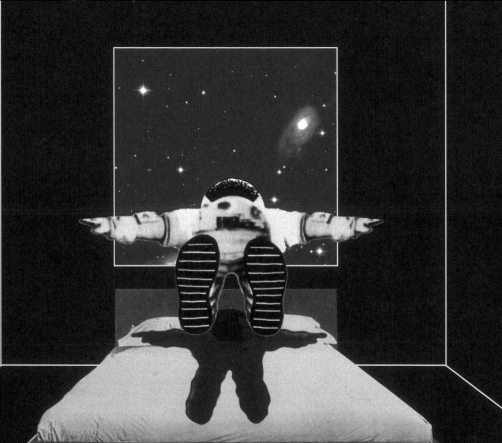

WAKING AFLOAT

Astronauts float when they come back to Earth. Or, at least that's what their brains trick them into believing. A few astronauts have reported the sensation of floating above their beds when they wake up. Astronaut Karen Nyberg recalled waking up and thinking she was weightless, hanging from her wall, looking down on her dresser from above.

EXERCISE

Astronauts need to exercise for two and a half hours each day while in space to combat losing bone density and muscle mass in microgravity. But exercising in space can be a sticky situation, in more ways than one. On the International Space Station, the exercise bike has no seat and no handlebars. There's no need for them. Astronauts simply strap into the foot pedals and get going, leaving their hands free to read a book or fiddle with their music playlists. Working up a sweat can be problematic, though, as it tends to stick to you, accumulating into large blobs, instead of floating away. Yet another good reason to always carry a towel in space.

BULL IN A CHINA SHOP

It's not easy being the new astronaut on a space station. It takes time to learn how to swim gracefully through the air. One shuttle pilot admitted that he created a floating mess of laptops and other devices in a wake behind him by flying too vigorously the first time he tried to fly from one room to another. "When you first turn up," he said, "you are like a bull in a china shop."

SPICE EXPLORATION

Meals in space can taste like cardboard, and not due to the unappealing nature of freeze-dried food. Lack of gravity causes astronauts' mucus to float around in their bodies rather than draining through their noses, creating congestion and often making them feel as though they have a head cold. As a result, astronauts suffer from a stuffy nose and lose some of their sense of smell, making everything taste bland. Stockpiles of Tabasco sauce, wasabi, horseradish, and Sriracha are sent into space frequently to keep astronauts' taste buds tingling.

OUTSIDE OF THE BUBBLE

Seeing the stars from space can make you feel like you've spent your entire life inside a bubble. Two-time space shuttle astronaut Mike Massimino described what it was like to look at the stars during a spacewalk:

It is kind of like looking at the Sun from the bottom of a swimming pool, versus looking at the Sun above the swimming pool. You are above that layer, so all of the stars, they don't twinkle. They are perfect points of light.

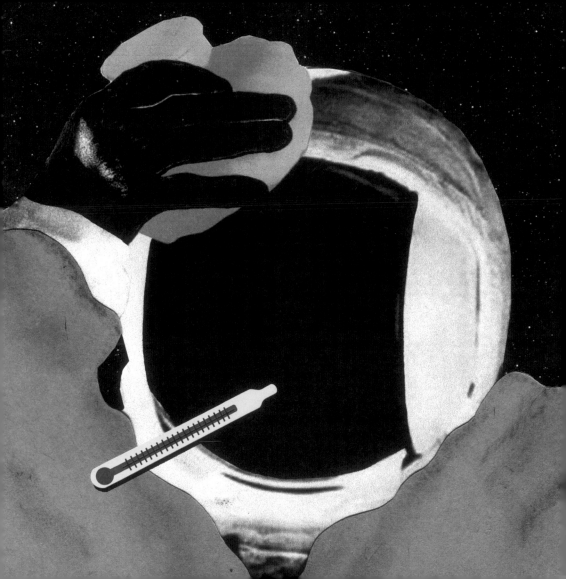

HEAD COLDS

Not surprisingly, getting a cold in space can be annoying. Without gravity to drain mucus, it stays floating inside your body. Ears and noses stay plugged up despite attempts to clear them. In the case of Apollo 7, the common cold caused all three astronauts to be irritated to the point of mutiny. With the threat of running out of tissues hanging over them and the real possibility of blowing out their eardrums, the astronauts refused to follow mission control safety instructions to wear their helmets for reentry.

LONG-TERM SPACE TRAVEL

The romanticism of being in space rubs off after a while. Belgian astronaut Frank de Winne explained:

If you are there for a week or two, you are basically on a high the whole time. It's not the same when you're there for six months. You need to manage your mood and motivation despite inevitable setbacks. Things that are difficult in the short term, such as not having a shower or any fresh fruit, become part of normal life. The things you really miss are close contact with your wife, your kids and your family and friends.

SOCK OVER THE DOORKNOB

Despite being isolated from most of the world while in space, there isn't a lot of privacy aboard space stations. This can be awkward for crews that have been all men and one woman. Some astronauts have developed systems not unlike the old dormitory sock-on-the-doorknob signal for when they need privacy. In space, a towel over the circular entrance to a docking compartment is the orbital DO NOT DISTURB sign.

UP ALL NIGHT

It's tempting to become an insomniac while in space. Anousheh Ansari, the first Iranian in space, found herself being too excited to sleep.

I tried not to sleep at night. I knew my time on the International Space Station was limited. I wanted to keep looking out the window at the Earth and the stars and take it all in. By the time I was ready to return back to Earth, I was so sleep deprived. I couldn't keep my eyes open.

Her crewmates couldn't believe she was falling asleep while they were preparing for reentry. It didn't last long, though. The G forces and orange glow of burning through the atmosphere will wake even the sleepiest of astronauts.

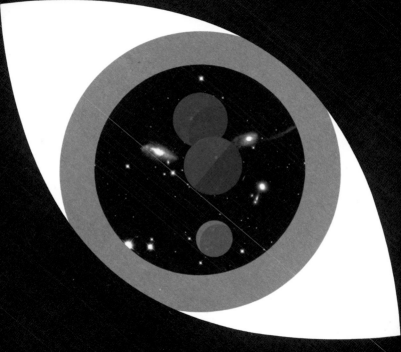

LIKE HOME

While aspects of being in space can become lackluster for some long-term space travelers, others find they have developed an entirely new appreciation for the experience. Astronaut Michael López-Alegría spent 215 consecutive days in space.

I was a little bit nervous about long duration missions being compared to a marathon. I was worried about being bored and wanting to get home. I found that couldn't be further from the truth! I really thrived being up there and having a pace that was more Earth-like. It's really living in space instead of just working in space. There's time to adjust, to reflect, to look out the window, to read a book, to write, to talk on the phone with people. If anything, you feel like you're more in space. It was a pleasant surprise. You feel closer to the environment. You feel like you're at home.

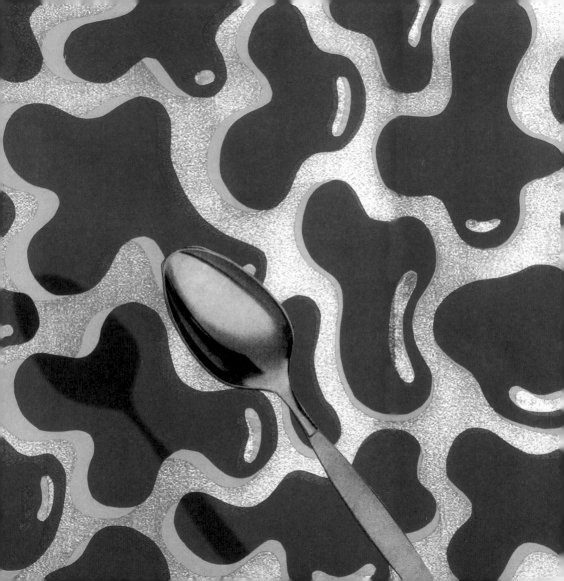

FOOD PLAY

All astronauts play with their food. Why use your hands when you can catch floating snacks with your mouth? Some foods are a bit tricky though, as astronaut Anousheh Ansari discovered:

When you open a bag of soft food like yogurt or soup, if you are not really, really careful, small yogurt bubbles or soup bubbles start floating around and then you can catch them with your spoon. But if you try to catch them too fast, one bubble hits your spoon and becomes ten smaller bubbles and now you have to catch ten of them!

THE OVERVIEW EFFECT

Many astronauts, when looking down on Earth over many orbits, experience an overwhelming shift in perception—a new awareness of the fragility and limited nature of the Earth. This experience has been dubbed "the overview effect." Astronaut Richard Garriott hadn't heard of the effect prior to experiencing it:

If you view the Earth from space, you can see results of natural and human forces at a macro scale. For example, the clouds that form over the Pacific tend to be very mathematical, very fractal. Whereas over the Atlantic, you get much more chaotic structures. You see the results of erosion by water and wind and the long scars of tectonic plate movement. You see crater impact zones all around the Earth. You see the impact of humanity all over the Earth. You see how every desert throughout the world is covered with roads and farms built through irrigation with deep ancient fossil water. You see roads crossing all the mountain ranges. You see dams on all the rivers. And finally after seeing all of that for a week, I saw the area where I grew up, and traveled by foot and car extensively. Suddenly my sense of the scale of the Earth went from being a large unknown to being both very finite and small.

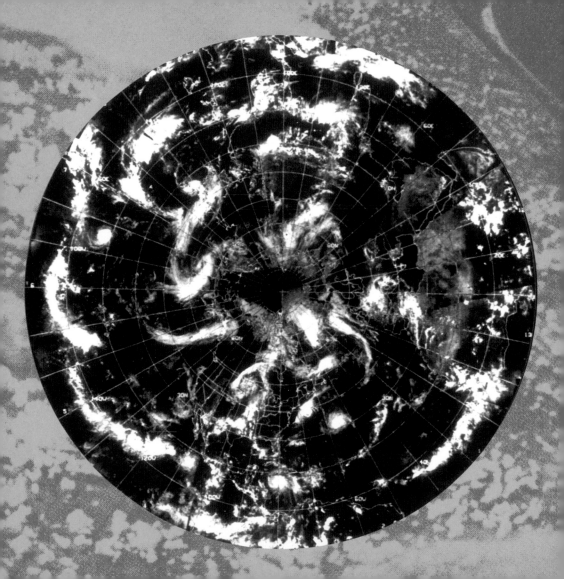

CLOUDS

Oceans and clouds cover the majority of the Earth. Three-time astronaut Sandy Magnus became fascinated with observing clouds from space:

Clouds from space seem to have personalities all their own. You get the big, angry thunderstorms, and the very orderly, polka-dotted clouds over the ocean, and then there are the puffy, heavy clouds. At sunset, you're going over thunderstorm clouds with pinks and reds and yellows like those you would see from the bottoms of clouds on Earth—but imagine seeing those colors on the tops of clouds. Sometimes you can even see the Moonlight reflected on the tops of clouds. You constantly have water and clouds as your companion.

YURI

The first human in space, Russian cosmonaut Yuri Gagarin, was acutely aware of what the beginning of human spaceflight represented in the context of human history. Prior to his launch in 1961, Gagarin explained the significance and responsibility he felt:

To be the first to enter the cosmos, to engage single handed in an unprecedented duel with nature—could anyone dream of anything greater than that? But immediately after that I thought of the tremendous responsibility I bore: to be the first to do what generations of people had dreamed of; to be the first to pave the way into space for mankind. This responsibility is not toward one person, not toward a few dozen, not toward a group. It is a responsibility toward all mankind—toward its present and its future.

CUBED OR TUBED

Space food in the early days of human spaceflight was either cubed or tubed. Meats were mashed into semisolid baby food–like consistencies and bottled into squeezable aluminum tubes. Other foods like cereal and cookies were crunched into cubes before being coated with either starch or a gelatin to reduce crumbs. Not looking forward to these square meals, astronaut John Young of Gemini 3 snuck a corned beef sandwich on board, but after he and fellow astronaut Gus Grissom took their first bites, crumbs went floating everywhere, risking clogging up vital spacecraft instruments. NASA was not pleased about the corned beef contraband.

SECOND BIRTH

When the meteor-like Russian Soyuz capsules land back on Earth from space, cosmonauts find it difficult to move their arms and legs, rendering them unable to climb out on their own. The search-and-rescue teams on the ground help by opening the capsule, cutting the cosmonauts out of their seatbelts, wrapping them in blankets, and carrying them out into the world.

OVERSCHEDULED

Every minute of every day is meticulously scheduled for astronauts. Between performing an extensive list of tasks and getting enough rest, there is little time to think about what it's like in space while being there. In 1973, feeling overscheduled by mission control, one group of astronauts aboard Skylab 4 went on strike. They turned off their radios, shut off communication with Earth, and spent the day thinking about the universe while gazing out the window. "We had been overscheduled," astronaut William Pogue wrote. "We were just hustling the whole day. The work could be tiresome and tedious, though the view was spectacular." Following the strike, NASA agreed to insert more down time into the astronauts' schedules for contemplating the cosmos.

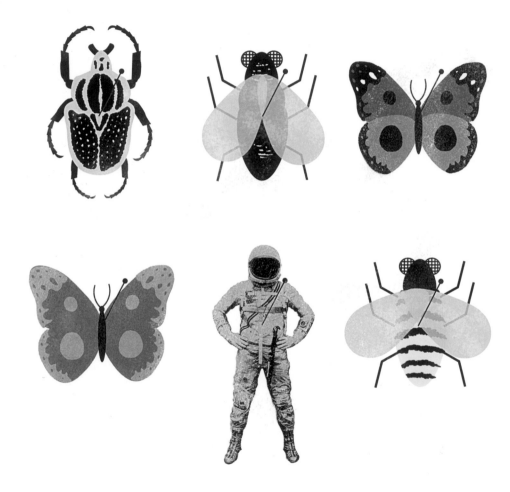

INSECTS IN SPACE

Spaceflight is all about learning how to "space float," whether you're a human or a housefly. Many winged insects have been sent into space, often with less-than-graceful results. Butterflies haphazardly bounce into barriers. Honeybees helplessly tumble inside enclosures. Houseflies find spaceflight troublesome enough to prefer walking on walls over flapping their wings. Moths, on the other hand, adapted to weightlessness; floating from place to place without a problem.

COOKIE CURRENCY

Onboard the first American space station, Skylab, astronauts created their own currency using sugar cookies. Freshly baked on Earth by the astronauts' nutritionist before takeoff, the cookies were a prized possession among the crew. If ever in need of a favor, they'd bribe a fellow crew member by giving up one of their coveted sugar cookies.

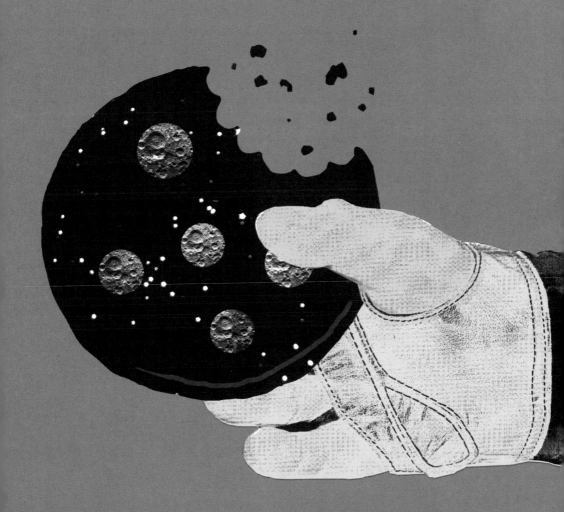

COSMIC FAIRIES

Every few minutes, astronauts are visited by what seem to be meddling cosmic fairies. Astronaut and noted astrophotographer Don Pettit explains:

In space I see things that are not there. Flashes in my eyes, like luminous dancing fairies, give a subtle display of light that is easy to overlook when I'm consumed by normal tasks. But in the dark confines of my sleep station, with the droopy eyelids of pending sleep, I see the flashing fairies. When a cosmic ray happens to pass through the retina it causes the rods and cones to fire, and you perceive a flash of light that is really not there. Free from the protection offered by the atmosphere, cosmic rays bombard us within [the International] Space Station, penetrating the hull almost as if it was not there. They zap everything inside, causing such mischief as locking up our laptop computers and knocking pixels out of whack in our cameras. The computers recover with a reboot; the cameras suffer permanent damage. After about a year, the images they produce look like they are covered with electronic snow.

AURORA

Many astronauts are mesmerized by seeing auroras from orbit. Canadian astronaut Chris Hadfield had the unique experience of flying over an aurora during his spacewalk:

They erupt out of the world and it's almost as if someone has put on this huge fantastic laser light show for thousands of miles. The colors, of course, with your naked eye are so much more vivid than just a camera. There are greens and reds and yellows and oranges and they poured up under my feet, just the ribbons and curtains of it—it was surreal to look at, driving through the Southern Lights.

ICE CREAM PARTY

Real astronaut ice cream is far more delightful than the freeze-dried pouches found in museums and gift shops. While it's true that the Apollo 7 mission had to endure the lackluster treat in 1968, later missions aboard *Skylab* and the space shuttle came equipped with freezers capable of carrying fresh ice cream. Astronauts aboard *Skylab* even held ice cream parties in space. The crew would gather around the window to watch the world go by while eating vanilla ice cream and strawberries.

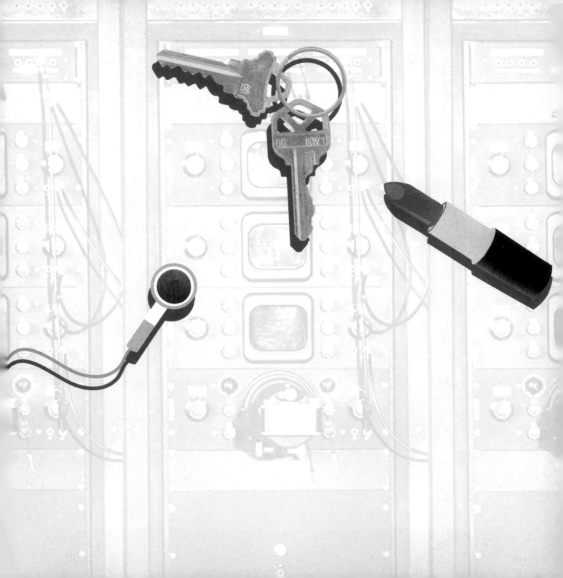

LOST AND FOUND

It's easy to lose things in space. Most space clothing has pockets with zippers for securing items, but it's hard to remember to zip them up every time you put something in them. Unsecured, items simply float out and away. First-time astronaut Anousheh Ansari lost her lip gloss this way. Worried that the floating item might do damage to the space station, she fessed up to one of her crewmates about it. Thankfully, she was told, the International Space Station has a large fan that sucks in anything floating around—essentially a lost and found box that they check once a week. Sure enough, the lip gloss was stuck in there, along with many other items that her crewmates had also accidentally left floating around.

SHRIMP COCKTAIL

Between the dehydrating, irradiating, and thermostabilizing the food intended for space endures, it's easy to see how most of it can be less than appetizing. In space, shrimp cocktail is considered a particular delicacy. Although freeze-dried, it's one of the few foods that retains its texture, while also packing a much-appreciated spicy punch of horseradish to taste buds that are less sensitive while in space. For decades, astronauts and cosmonauts alike have celebrated shrimp cocktail as their favorite space cuisine, with some so obsessed with it that they would eat it at every meal—breakfast, lunch, and dinner—for weeks.

JULY

MON	TUE	WED	THU	FRI	SAT	SUN
			1	2	3	4
5	6	7	8	9	10	11
12	13	14	15	16	17	18
19	20	21	22	23	24	25
26	27	28	29	30	31	

SATELLITES AND METEORS

Astronauts orbiting the Earth are surrounded by all sorts of space objects and vistas. Astronaut Don Pettit observed of his time aboard the International Space Station:

You catch an occasional meteor while looking down. You see stars and planets and our galaxy on edge. . . . You see space junk orbiting nearby. Sometimes it flickers due to an irregularity catching light as it rotates. . . . You observe other satellites, some in equatorial orbits, some in polar. You notice some satellites above your orbit only visible while looking away from Earth. Like a short-lived magnesium flare, some satellites flash brilliantly for a few seconds and fade into oblivion.

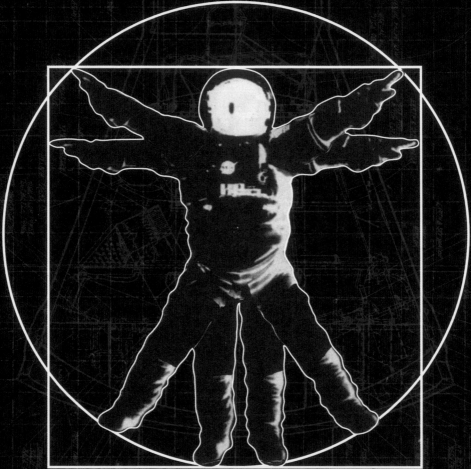

NOBODY KNEW

No one knew what it was like in space before sending a human up there. Doctors were concerned that astronauts' eyeballs might float around in their sockets, causing blindness, or that the astronauts might starve if unable to swallow in weightlessness. In what may have been an overabundance of caution, early astronauts were subjected to many, many physical and psychological tests that might be considered embarrassing at best, and extremely uncomfortable at worst. When asked what test was the most difficult, Mercury astronaut John Glenn replied, "It is rather difficult to pick one because if you figure how many openings there are on a human body, and how far you can go into any one of them—you answer which one would be the toughest for you."

MEASURING EVERYTHING

Everything about the experience of early astronauts in orbit was meticulously measured, from their vital signs to their food intake to their waste output. Onboard *Skylab*, they measured every meal that went in and saved every meal that went out for analysis on the ground. This presented a problem for astronaut William Pogue who threw up while in space. Not wanting mission control to know he had been ill, he threw his bag of puke out into the airlock. His sneaky strategy almost worked—until later he realized that he had discussed the scheme over the intercom with a fellow astronaut for everyone to hear.

SCARING YOURSELF

It takes a while to get used to the unusual things that happen in space. For instance, when your body is relaxed enough to fall asleep, your arms end up floating around instead of remaining by your side. Astronaut Vance Brand scared himself silly by forgetting this. While some of the crew bundled up in sleeping bags, he simply tethered himself to a handrail with a clip. After drifting off to sleep, he awoke in the middle of the night to see something dangling right in front of his face that scared the hell out of him: his own hands.

BEING HUMAN

Spare time is a rare commodity for astronauts performing highly-coordinated space missions. Apollo astronaut Rusty Schweickart recalled what it was like to even find five spare minutes while on a spacewalk:

I looked at the Earth, and I just said, "My job right now is to just be a human being, just be a person." And I just stopped being an astronaut. There I was, a human being in space, saying, "Absorb this. Just soak this up. Just let it all come in" No defenses, just ultimately vulnerability. And I just became a human being in space, looking at this beautiful planet . . . it was this huge philosophical big hit. . . . So that five minutes was a very special five minutes.

EARTH AS A WORK OF ART

Earth's ocean currents and seabeds are like an art gallery for astronauts as they look down from space. Some astronauts fall in love with distinct works of Earth art. A frequenter to the International Space Station, astronaut Sandy Magnus reflected on her favorite vista:

Absolutely hands down one of the most beautiful places to see from space is the Caribbean. You see an entire rainbow of blue. From the light emerald green to the green blue to the blue green to the aqua marine to the slowly increasingly darker shades of blue down to the really deep colors that come with the depths of a really deep ocean. You can see all that at one time from space. It's very curvy, it's not harsh geometric lines. Its swirls and whirls and all kinds of wavy lines. It looks like a piece of modern art.

IN YOUR OWN BACKYARD

Most astronauts spend relatively little time in space, serving on missions that only last for a week or two. For those who go to space for months, the experience is very different. Astronaut Ron Garan reflected on the difference between his short- and long-duration stays on the International Space Station:

You learn how to really exist in this new environment. When you look at the Earth on a short duration mission you're seeing one snapshot of the life on the planet, but when you're up there for months, you really get this sense that you're watching a living breathing organism. On my short duration mission, I would look down during a spacewalk and see this absolutely beautiful part of the planet and say "Ooo, where are we?" On my long duration mission, I had four or five months of looking out the window and when I went on a spacewalk I really felt like I was going into my own backyard. I wouldn't have to ask anybody where we were.

SPACEWALK

While performing a spacewalk is an exciting experience, it is also a very serious operation that is meticulously scripted for astronauts. The only time astronauts might get a chance to look around at where they are is when there's a glitch in equipment and they get a few spare minutes while someone makes a repair. Astronaut Chris Hadfield found an opportunity to look around during one of his spacewalks:

The contrast of your body and your mind inside . . . essentially a one-person spaceship, which is your spacesuit, where you're holding on for dear life to the shuttle or the station with one hand, and you are inexplicably in between what is just a pouring glory of the world roaring by, silently next to you—just the kaleidoscope of it, it takes up your whole mind. It's like the most beautiful thing you've ever seen just screaming at you on the right side, and when you look left, it's the whole bottomless black of the universe and it goes in all directions. It's like a huge yawning endlessness on your left side and you're in between those two things and trying to rationalize it to yourself and trying to get some work done.

SUNRISE

In orbit, astronauts experience the Sun rising over the Earth every ninety minutes. Astronaut Mike Mullane observed of his 200 sunrises during 365 hours in space:

I think if there's one thing that you could truly say is the most beautiful sight you can possibly see as a human, it is watching sunrise over the Earth, because imagine, you're looking at blackness out the window, black Earth, black space, and then as the Sun comes up, the atmosphere acts as a prism, and it splits the light into the component colors. It splits the white light of the Sun into the component colors, so you get this rainbow effect, and it starts with this deep indigo eyelash, just defining the horizon, and then as the Sun rises higher, you get these reds and oranges and blues in this rainbow. . . . You never got tired of looking at those.

LOCAL COLOR

Surrounded by the same environment each day, astronauts' senses become finely attuned to subtle changes. Some can tell what parts of the world they're flying over by the change in color reflected into the cabin. Others know which ocean they're looking at by the shapes of the clouds. A few can identify countries simply by the geometrical shapes of their farmland. German astronaut Alexander Gerst tweeted from space, "When light from the Cupola tints #ISS orange inside, I can tell we're over Africa without even looking out the window."

SO YOU *CAN* BURP IN SPACE

While the lack of gravity in orbit makes burping essentially synonymous with vomiting, astronaut Jim Newman came up with a trick for separating the two. He found that by pushing off a wall, he could create a force in lieu of gravity that kept his food down in his stomach, giving him a brief chance at expelling gas without consequence. He dubbed it the "push and burp."

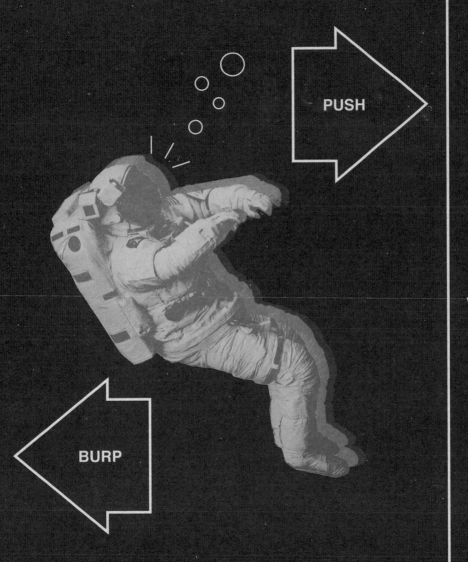

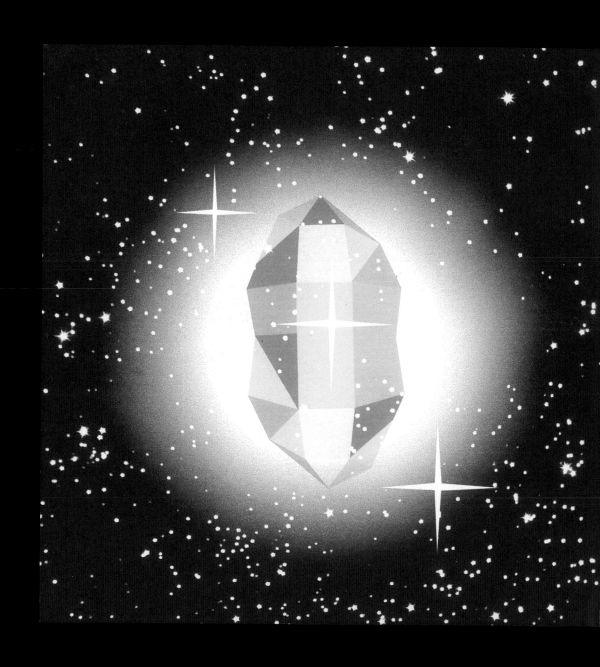

CRYSTAL SPECTACLE

While bodily fluids are normally something most of us prefer kept out of sight, in space they can be a must-see spectacle. Excess water from the fuel cells and urine expelled into the vacuum of space via a liquid waste venting system quickly boils and then freezes, creating little ice crystals. Astronaut Jim Newman described the unexpected beauty of this sight from aboard the space shuttle: "It is astonishingly beautiful to see this big stream of ice particles going out, being illuminated by the Sun and reflecting the light. It's just gorgeous."

OUR ORNAMENT

One of the most significant sights ever to be witnessed by humans was when the crew of the 1968 Apollo 8 mission looked out the window to see the Earth rise over the Moon. "To see this very delicate, colorful orb which to me looked like a Christmas tree ornament coming up over this very stark, ugly lunar landscape . . . That one view is sunk in my head," recalled astronaut Bill Anders. "Here was this orb looking like a Christmas tree ornament, very fragile, not an infinite expanse of granite . . . and seemingly of a physical insignificance and yet it was our home."

Acknowledgments

To Matt for sharing life, the universe, and everything with me.

Thanks to the amazing astronauts who graciously shared their time and truly inspiring stories with me: Anousheh Ansari, Brian Duffy, Ron Garan, Richard Garriott, Greg H. Johnson, Michael López-Alegría, Ed Lu, Sandy Magnus, Pam Melroy, Jim Newman, and Bryan O'Connor. You all put a sparkle in my eye each time I think back to our conversations about space travel. Thanks also to my astronaut wrangler friends Mary Lynne Dittmar and Chris Gerty. Additionally, I'd like to send my thanks and appreciation to the many astronauts (past and present) mentioned and quoted in this book who frequently share their unique experiences with the world. Finally, thanks to my husband Matt Biddulph, who shared every laugh and disgusted look with me along the way from the first story to the last, and became a total space geek in the process.

About the Author

Ariel Waldman is a hacker of space exploration and a maker of "massively multiplayer science." She instigates unusual collaborations that spark clever creations for science and space exploration.

Ariel is the founder of Spacehack.org, a directory of ways to participate in space exploration, and she is the global director of Science Hack Day, an event that brings scientists, technologists, designers, and enthusiastic people together to see what they can create in one weekend. She sits on the council for NASA Innovative Advanced Concepts, a program that nurtures radical, sci-fi–esque ideas that could transform future space missions. Ariel has also served as an appointed National Academy of Sciences committee member on a congressionally-requested study on the future of human spaceflight.

Previously, Ariel worked at NASA's CoLab program whose mission was to connect communities inside and outside NASA to collaborate. She has keynoted DARPA's 100 Year Starship Symposium and O'Reilly's Open Source Convention, as well as appeared on the SyFy channel as part of their Let's Imagine Greater campaign. In 2013, Ariel received an honor from the White House for being a Champion of Change in citizen science.

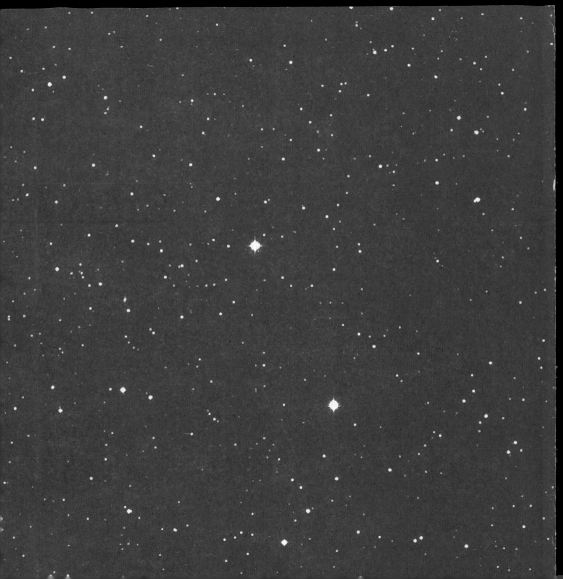